TIBETAN MEDICINAL PLANTS

by the same author

Fundamentals of Tibetan Medicine
Handbook of Traditional Tibetan Drugs

Tibetan Medicinal Plants

Tsewang J. Tsarong

Tibetan Medical Publications
Kalimpong, WB, India

First Edition 1994

TIBETAN MEDICINAL PLANTS

© 1994 by Tibetan Medical Publications

All rights reserved. No part of this work covered by the copyright hereon may be reproduced or used in any form or by any means—graphics, electronic, or mechanical, including photocopying, recording, taping, or information storage and retrieval systems—without written permission of the publisher.

Printed in India

ISBN 81-900489-0-2

Published by Tibetan Medical Publications
Penrhyn House, W. Rickshaw Road
Kalimpong - 734 301
(District Darjeeling)
W. Bengal, INDIA

Typeset & Printed at
The Services Press, 40-A, Khan Market,
New Delhi - 110 003 India

In Memory of
Dasang Dadul Tsarong
(1888 - 1959)

Acknowledgement

To all those who helped me during the years of preparation, I fully extend my grateful acknowledgement with special thanks to:

Amchis Jamyang Tashi (late), Yeshi Sonam (late), Lobsang Chöphel, Tempa Chöphel, Lobsang Tenzin, Dawa, Kunga Gyurmed Nyarongshar, Tsewang, Dhoptra Amchi, Lobsang Tashi, Gyurmed Norbu, and Kalsang for the proper identification of indigenous medicinal plants.

Paljor Publications for the Tibetan fonts.

Professor Heinrich Harrer for the use of his slides and who first introduced me to the fun and pleasure of trekking and photographing plants in the majestic Himalayas.

Last, but not least, to Senator Gunther Klinge of Munich, without whose kind interest and support, this publication would not have been possible.

Contents

Preface	8
Medicinal Plants	11
Notes	109
Glossary	113
References	118
Index of Latin Names	121
Index of Tibetan Phonetics	124
Index of Tibetan Names	127

Preface

High in the Land of Snows, amidst melting snow-beds, steep screes, and glacial moraines grow some of the loveliest and most colourful flowers in the world. Many of these wild and exotic plants were used for centuries as ritual offerings and healing drugs by the lama-physicians of Tibet. These healers (popularly known as *amchis* or *menpas*), through painstaking trial, error, and observation, have identified these plants and documented their therapeutic action and uses in herbals that date as far back as 8th century C.E.

The aim of this book is to draw attention to the above herbal legacy and to generate an interest in one of the most fascinating branches of Tibetan medicine.

The photos in this book are the result of innumerable treks I undertook during the past decade or so, to various regions of the Himalayas. The areas I covered are Dzongri in Sikkim, Hemkund and the Valley of the Flowers in the Garhwal Hills of Uttar Pradesh, Kanji in the Zangskar region of Ladakh, Mari/Rothang Pass in Himachal Pradesh, and Solu Khumbhu in Nepal.

The specific information on the pharmacodynamics of these plants is taken from the *rGyud-bZhi* and the *Shel-gong Shel-preng*. The *rGyud-bZhi* is the most important Tibetan medical text and provides the names and therapeutic action of 111 different plants. Moreover, it contains valuable information on methods of plant collection, storage, detoxification, duration of potency, and pharmacological techniques to make them easier to ingest and assimilate.

On the other hand, the *Shel-gong Shel-preng*, written in 1727 C.E. by Geshe Tenzin Phuntsog, is the best known Tibetan pharmacopoeia. It lists 2294 drugs from plants, animal extracts, rocks, salts, and minerals. The plant section provides different names and therapeutic action of 312 herbs.

It is my sincere hope that this book will revive an interest in this field of traditional medicine and encourage students and researchers of Tibetan medicine to complete the full documentation of these wonderful plants that have played such a vital role in the healthcare of Tibetans for so many centuries.

T.J. Tsarong
Kalimpong, W. Bengal
June, 1994

Medicinal Plants

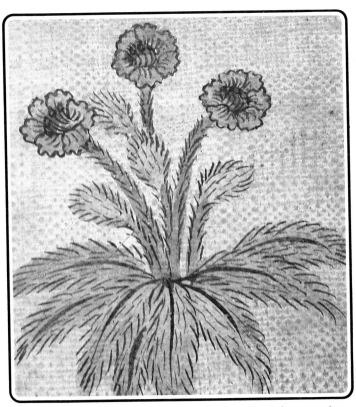

Tsayr-ngön - *Meconopsis* sp. from an old Tibetan herbal manuscript.

བ་ལུ།

Bah-lhoo[1] - **Rhododendron anthopogon var. hypenanthum**
Taste: Sweet, bitter, and astringent
Potency[2] **:** Promotes heat
Action and Use[3]**:** Promotes heat, digestive, antitussive. Used against lack of *Me-drod*, lack of appetite, coughing, and various skin disorders.
Parts Used: Stem and leaves.

ཨ་ཤ་ཀ།

Bha-shah-kah - **Phlogacanthus pubinervius**
Taste: Bitter
Potency: Cooling
Action and Use: Febrifuge, anti-inflammatory, dries up *Ngan-khrag*. Used against proliferation of *Ngan-krag*, inflammation of blood, liver, and 'hot' *mKhris-pa*.
Parts Used: Stem, leaves, and flower.

བོང་ང་དཀར་པོ།

Bhong-ngah Kar-poh - **Aconitum heterophyllum**
Taste: Bitter
Potency: Cooling
Action and Use: Anti-inflammatory, analgesic, febrifuge (specially from poisoning), cholagogue. Used against poisoning from scorpion or snake bite, fever from contagious diseases, 'hot' *mKhris-pa,* and inflammation of the intestines.
Parts Used: Tuber.

བོང་ང་ནག་པོ།

Bhong-ngah Nhuck-poh[4] - **Aconitum balfourii**
Taste: Acrid and sweet
Potency: Promotes heat and very poisonous
Action and Use: Anti-inflammatory, analgesic, vermifuge, antirheumatic, dries up serous fluids, controls *rLung*. Used against all types of pain and inflammations from gout or arthritis; all disorders due to worms/microorganisms, *sNying-rLung*, amnesia, loss of bodily heat, leprosy, and paralysis.
Parts Used: Tuber.

ལྕམ་པ།

Cham-pah - **Malva verticillata**
Taste: Sweet and astringent
Potency: Promotes heat
Action and Use: Diuretic, antidiarrheal, and heals renal disorders. Used against retention of fluids, frequent thirst, diarrhea, and various *Tza* disorders.
Parts Used: Seeds.

ཞུ་ཚ།

***Chay-tsah* - Runanculis acris**
Taste: Acrid
Potency: Promotes heat
Action and Use: Promotes heat, dissolves tumors, and draws out serous fluids. Used against lack of *Me-drod*, formation of tumors, disorders brought about by rotting sores/wounds, *Gag-Lhog,* and *dMu-chu.*
Parts Used: Flowers.

བཅན་ཟིག་ཏག་ངུ།

***Chen-zeek Tagh-nghoo* - Saxifraga sp.**
Taste: Sweet and astringent
Potency: Neutral
Action and Use: Tonic, promotes vigor, and balances the three processes. Used against loss of bodily vigor, lack of essential blood (*Zung-khrag*), loss of clarity of the senses, and imbalance of the three primary processes of *rLung, mKhris-pa,* and *Bad-kan.*
Parts Used: Entire plant.

ཆུ་མ་རྩི།   Prof. H. Harrer

Choo-mah-tzee - **Rheum nobile**
Taste: Sour
Potency: Promotes heat
Action and Use: Laxative, diuretic, antiemetic. Used against retention of body fluids, swelling, and fullness of abdomen.
Parts Used: Stem and flowers.

ཆུ་རུག་སྦལ་ལག

Choo-rhook Bell-luck - **Halerpestes sarmentosa**
Taste: Sweet
Potency: Neutral
Action and Use: Febrifuge, anti-inflammatory, diuretic. Used against inflammation of ureters, pain in joints due to arthritis, gout, and retention of bodily fluids.
Parts Used: Stem and leaves.

ཤུགས་ཏིགས་དཀར་པོ།

Chuck-theek Kar-poh - **Swertia petiolata**
Taste: Bitter
Potency: Extremely cooling
Action and Use: Anti-inflammatory, febrifuge, liver tonic, fever from serous disorders. Used against scleritis, inflammation of liver and high blood pressure from *mKhris-pa*, inflammation of stomach and kidneys from excessive 'hot' diet and lifestyle.[5]
Parts Used: Entire plant.

ཤུགས་ཏིགས་ནག་པོ།

Chuck-theek Nhuck-poh - **Gentiana sp.**
Taste: Bitter
Potency: Very cooling
Action and Use: Febrifuge, cholagogue, anti-inflammatory. Used against jaundice, inflammation of sclera, liver, and kidneys.
Parts Used: Entire plant.

ཞུགས་ཏིག་ར་མགོ་མ།

Chuck-theek Rah-goh-mah - **Halenia elliptica**
Taste: Bitter
Potency: Cooling
Action and Use: Febrifuge, cholagogue. Used against 'hot' *mKhris-pa*, inflammation of liver, stomach, and fever from contagious diseases.
Parts Used: Entire plant.

ཤྱང་ཚེར་ནག་པོ།

Ch'ang-tser Nhuck-poh - **Morina longifolia**
Taste: Sweet and astringent
Potency: Promotes heat
Action and Use: Emetic, stomachic, digestive. Used against stomachal disorders such as indigestion giving rise to emesis and nausea.
Parts Used: Stem, leaves, and flowers.

ཐེ་ག

Day-gah - **Thlaspi arvense**
Taste: Acrid
Potency: Cooling
Action and Use: Febrifuge, anti-inflammatory. Used against pus in the lungs, renal inflammation, appendicitis, seminal and vaginal discharge.
Parts Used: Seeds.

འབྲི་ཏ་ས་འཛིན།

Dee-tah Sah-zheen - **Fragaria sp.**
Taste: Slightly sweetish
Potency: Cooling
Action and Use: Febrifuge, hemostatic, dries up serous fluids. Used against spreading of putrid blood and pus in upper chest regions specially in the pulmonary region, and fever of *rTza*.
Parts Used: Entire plant.

ཧུ་ལིས།

***Dhah-lhee*[6]** - **Rhododendron anthopogon var. hypenanthum**
Taste: Sweet
Potency: Neutral
Action and Use: Febrifuge, antitussive, tonic. Used against swelling of body due to *sKya-rBab*, inflammations, disorder of the lungs, and general weakening of body. Moreover, it is used when water and locality are not agreeable due to change of environment.
Parts Used: Flowers.

དུ་བ།

***Dah-wah* - Arisaema sp.**
Taste: Acrid
Potency: Promotes heat
Action and Use: Anti-inflammatory, antibacterial, anthelmintic, and heals rotting tissues. Used against stomachal/intestinal pains from worm infestations, inflammatory swellings, and tumors.
Parts Used: Tuber.

དན་རོག

Dhen-rock - **Ricinus communis**
Taste: Acrid, bitter, and sweet
Potency: Promotes heat
Action and Use: Purgative, expellant. Used against *Bad-kan* disorders such as indigestion and also as an expellant or a purgative for a combination of either two or three of the primary processes of *rLung*, *mKhris-pa*, and *Bad-kan*..
Parts Used: Seeds.

དྲེས་མ།

Drhay-mah - **Iris kemaonensis**
Taste: Acrid
Potency: Cooling
Action and Use: Analgesic, increases eyesight. Used against tinnitis, pain in the ears, and weakening of eyesight.
Parts Used: Flowers.

ཇྲེས་མའི་གེ་སར།

***Drhay-mare Gay-sahr* - Iris kemaonensis**
Taste: Acrid
Potency: Promotes heat
Action and Use: Anthelmintic, analgesic, vermifuge. Used against colic pain due to intestinal worms, hot and cold disorders of the stomach/intestines, and pain below the neck/shoulders.
Parts Used: Seeds.

འབྲི་མོག

Drhee-mock **- Onosma sp.**
Taste: Sweet
Potency: Cooling
Action and Use: Febrifuge, heals pulmonary disorders, dries up *Ngan-khrag*. Used against inflammation of the lungs, bursting of air passages in the lungs, and spreading of *Ngan-khrag* .
Parts Used: Roots.

སྨྱུག་ཞགས།

Drhül-shuck - **Cuscuta eropaea var. indica**
Taste: Bitter, acrid, sweet
Potency: Promotes heat
Action and Use: Renal, vertebral, hepatic tonic, increases semen, aphrodisiac. Used for pain in the waist, limbs, vaginal/seminal discharge, polyurea, tinnitis, and blurred vision.
Parts Used: Entire plant.

དབྱི་མོང་དཀར་པོ།

***E-mong Kar-poh* - Clematis sp.**
Taste: Acrid and sweet
Potency: Promotes heat
Action and Use: Promotes heat, dissolves tumors, digestive. Used against lack of *Me-drod*, indigestion, and tumors.
Parts Used: Stem and flowers.

དབྱི་མོང་ནག་པོ།

E-mong Nhuck-poh - **Clematis tibetana**
Taste: Acrid and Sweet
Potency: Promotes heat
Action and Use: Promotes stomachal heat, destroys 'cold' tumors.[7] Used against skin irritations/itchings, lack of *Me-drod*, and tumors.
Parts Used: Stem and flowers.

Gah-kyah - **Hedychium spicatum**
Taste: Acrid
Potency: Promotes heat
Action and Use: Promotes heat, vasodilator, digestive, stomachic, heals *Bad-rLung*,[8] Used against loss of *Me-drod*, indigestion, *Bad-rLung*, and poor circulation due to thickening of blood.
Parts Used: Rhizome.

སྲ་ཏིག

Gah-theek - **Androsace sarmentosa**
Taste: Bitter
Potency: Cooling and coarsening
Action and Use: Resolutive, dries up serous fluids. Used against disorders from tumors, *sKya-rBah*, inflammation of fluids and other serous fluid disorders.
Parts Used: Entire plant.

གཱ་ཤྲ།

Gah-trah - **Rubus ellipticus**
Taste: Sweet and sour
Potency: Promotes heat
Action and Use: Renal tonic, vaginal/seminal discharge, antidiuretic. Used against weakening of the senses, vaginal/seminal discharge, polyuria, and micturation during sleep.
Parts Used: Inner bark.

ཀོ་སྙོད།

Goh-nyöd - **Carum carvi**
Taste: Acrid
Potency: Promotes heat
Action and Use: Blocks *rLung* openings,[9] febrifuge, strengthens vision, digestive. Used against 'hot' *rLung*, failing vision, and loss of appetite.
Parts Used: Seeds.

ཧ་ལོ།

Hah-loh - **Althea rosea**
Taste: Sweet and acrid
Potency: Neutral
Action and Use: Anti-inflammatory, aperitive, stops vaginal/seminal discharge. Used against inflammation of kidneys/womb, vaginal/seminal discharge, and the roots are used for loss of appetite.
Parts Used: Flower and roots.

ཧོང་ལེན།

Hong-lhen - **Lagotis sp.**
Taste: Bitter
Potency: Cooling
Action and Use: Febrifuge, anti-inflammatory, dries up *Ngan-khrag*. Used against inflammation of liver, lungs, intestines, and spreading of *Ngan-khrag*.
Parts Used: Entire plant.

བྱ་རྐང་།

Jah-k'ang - **Delphinium sp.**
Taste: Bitter
Potency: Cooling
Action and Use: Febrifuge, antidiarrheal, heals wounds. Used against diarrhea from inflammatory infections, and lice infestation.
Parts used: Flowers.

བྱར་པན་ཆུ་རྩེ།

Jahr-phen Chu-tzey - **Epilobium latifolium**
Taste: Bitter
Potency: Cooling
Action and Use: Antitoxin, febrifuge, analgesic, stops itching, antirheumatic, anti-inflammatory. Used against fever and inflammation from *'Bam*, rheumatism, and itching pimples.
Parts Used: Entire plant.

ཀ་ཀོ་ལ།

Kah-koh-lah - **Amomum sabulatum**
Taste: Acrid
Potency: Promotes heat and coarsening
Action and Use: Stomachic, digestive, carminative. Used against loss of stomachal/splenic heat, indigestion, swelling, and fullness of abdominal region.
Parts Used: Fruit.

གནད་ཀ་རི།

***Khen-drah-kah-rhee* - Rubus sp.**
Taste: Sweet and astringent
Potency: Neutral
Action and Use: Balances the three processes of *rLung*, *mKhris-pa*, and *Bad-kan*, antitussive, febrifuge, anti-inflammatory, antitoxin. Used against the common cold, unripened fever from contagious diseases,[10] cough, and pulmonary disorders.
Parts Used: Inner bark.

ཁུར་མང་།

Khoor-m'ang - **Taraxacum tibeticum**
Taste: Bitter
Potency: Cooling
Action and Use: Febrifuge, anti-inflammatory, anthelmintic, heals *sMug-po*. Used against stomachal disorders such as *sMug-po*, and pain in stomach/intestines due to intestinal worms.
Parts Used: Entire plant.

གོན་པ་སྒབ་སྐྱེས།

Khön-pah Ghab-keh - **Saussurea roylei**
Taste: Bitter
Potency: Cooling
Action and Use: Anti-inflammatory, vasoconstrictor, antitoxin, emollient, hemostatic. Used against wounds, excessive bleeding, and meat poisoning.
Parts Used: Entire plant.

སྐྱུ་རུ་ར།

Kyou-rhoo-rah - **Emblica officinalis**
Taste: Sour
Potency: Cooling
Action and Use: Balances the three processes of *rLung*, *mKhris-pa*, and *Bad-kan*; removes *Ngan-khrag*, febrifuge, anti-inflammatory, antidiuretic. Used against *Bad-mKhris*, proliferation of *Ngan-khrag*, inflammation of blood, 'hot' *mKhris-pa*, polyuria, and loss of hair.
Parts Used: Fruit.

སྲེ་ངེས།

***Ley-trey* - Tinospora cordifolia**
Taste: Sweet, bitter, and acrid
Potency: Neutral
Action and Use: Balances *rLung*, *mKhris-pa*, and *Bad-kan*; heals 'hot' *rLung*, anti-inflammatory, analgesic, and dries up serous fluids. Used against imbalance of the three processes, 'hot' *rLung*,[11] unripened/contagious fever,[12] inflammation/swelling of joints due to arthritis, gout, and serous disorders. Moreover it is use against all types of pain and 'chronic fever.'[13]
Parts Used: Stem.

ཨེ་ག་དུམ།

Lhee-gah-dhoor - **Geranium pratense**
Taste: Acrid and sweet
Potency: Cooling
Action and Use: Febrifuge, analgesic, anti-inflammatory. Used against fever from influenza, inflammation of lungs, *rTza*, pain and swelling of the limbs.
Parts Used: Roots.

གླུ་བདུད་རྡོ་རྗེ།

***Lhoo-dü Dhor-jay* - Codonopsis nervosa**
Taste: Sweet and astringent
Potency: Cooling
Action and Use: Anti-inflammatory, analgesic, dries up serous fluids. Used against pain and swelling of joints due to gout/arthritis, stiffening of ureters, and paralysis due to cerebral ischemia.
Parts Used: Flowers.

ཨུག་མིག

***Lhook-meek* - Aster diplostephioides**
Taste: Bitter
Potency: Cooling
Action and Use: Antitoxin, febrifuge, tonic, hemostatic. Used against infectious fevers, influenza, nose-bleeding, poisoning, sores from environmental poisoning, and inability to stretch or contract the limbs.
Parts Used: Flowers.

ལུག་རུ་དཀར་པོ།

Lhook-rhoo Kar-poh - **Pedicularis bicornuta**
Taste: Bitter
Potency: Cooling
Action and Use: Facilitates controlling action of *rLung Thur-sel*. Used against vaginal and seminal discharges.
Parts Used: Flowers.

ལུག་རུ་སྨུག་པོ།

Lhook-rhoo Mhook-poh - **Pedicularis oliveriana**
Taste: Bitter
Potency: Cooling
Action and Use: Antitoxin, anti-inflammatory, febrifuge, anti-diarrhea. Used against poisoning, inflammation of the stomach/intestines, diarrhea, stomachal *sMug-po* disorders, and heals obstinate sores and wounds.
Parts Used: Flowers.

ཀྱུག་ཏུ་སེར་པོ།

Lhook-rhoo Sayr-poh - **Pedicularis longiflora var. tubiformis**
Taste: Bitter and astringent
Potency: Cooling
Action and Use: Febrifuge, diuretic, facilitates controlling action of *rLung Thur-sel*, removes excess accumulation of body fluids. Used against inflammation of liver/gall bladder, seminal/vaginal discharges, edema, and disorders associated with alcoholism.
Parts Used: Entire plant.

ལང་སྐྱ།

L'ang-nah - Pedicularis pyramidata
Taste: Sweet and astringent
Potency: Promotes heat
Action and Use: Removes excess body fluids, diuretic, analgesic, and regulates respiration. Used against retention or accumulation of fluids, difficulty in micturation, breathlessness, inflammation of bone/marrow, pain from inflammation and accumulation of serous fluids.
Parts Used: Entire plant.

མ་ནུ།

Mah-nhoo - **Inula racemosa**
Taste: Sweet, bitter, and acrid
Potency: Neutral
Action and Use: *Khrag-rLung* and *sMug-po* regulator, analgesic. Used against *Khrag-rLung*, *sMug-po*, contagious fevers that have not fully ripened,[14] pain in upper body specially between the neck and shoulders.
Parts Used: Rhizome.

མེ་ཏོག་གངས་ལྷ།

May-thock G'ang-lhah - **Saussurea graminifolia**
Taste: Sour and sweet
Potency: Promotes heat
Action and Use: Antitussive, purifies and increases blood, aphrodisiac, promotes heat in the womb, emmenagogue. Used against coughing due to loss of potency of the spleen, irregular menses, seminal/vaginal discharge, excessive bleeding from the womb, and pain of the waist due to loss of renal potency.
Parts Used: Entire plant.

མིང་ཅན་ནག་པོ།

Prof. H. Harrer

Ming-chen Nhuck-poh - **Cremanthodium sp.**
Taste: Bitter
Potency: Cooling
Action and Use: Anti-inflammatory, febrifuge specially of fever from poisoning. Used against inflammatory disorders such as *Gag-Lhog*, and influenza.
Parts Used: Entire plant.

མིང་ཅན་སེར་པོ།

Ming-chen Sayr-poh - **Cremanthodium sp.**
Taste: Bitter
Potency: Cooling
Action and Use: Anti-inflammatory, analgesic, balances *Khrag-rLung*. Used against pain in upper bodily regions due to *Khrag-rLung*, and contagious febrile and inflammatory disorders such as *Gag-Lhog*.
Parts Used: Entire plant.

ཉིམ་པ།

Neem-pah - **Azadirachta indica**
Taste: Bitter
Potency: Neutral
Action and Use: Febrifuge, carminative, clears mouth odor, promotes hair growth. Used against lack of appetite, excessive thirst, bad breath, skin disorders such as *Me-dBal*, and loss of hair.
Parts Used: Stem and leaves.

ནད་མ་གཡུ་ལོ།

Neh-mah Yhoo-loh - **Hackelia uncinata**
Taste: Sweet and bitter
Potency: Neutral
Action and Use: Subsides coughing, expectorant, antitussive, heals wounds, destroys tumors. Used against coughing, tumors in the womb, sores, wounds, and swelling of body due to *sKya-rBab*.
Parts Used: Flowers.

ཡོ་བྲོ་འང་རྩེ།

Ngoh Droh-sh'ang-tzey - **Orobanche sp.**
Taste: Sweet
Potency: Promotes heat
Action and Use: Renal tonic, aphrodisiac, increases power of vetebrae. Used against pain in the waist or legs, lack of sexual appetite, and general weakening of the senses.
Parts Used: Entire plant.

Ngoh-trheen - **Thalictrum sp.**
Taste: Bitter
Potency: Very cooling
Action and Use: Febrifuge, antitoxin, antirheumatic, anti-inflammatory, antidiarrheal. Used against inflammation of scalera, inflammation of liver from infections, boils/pimples, and diarrhea.
Parts Used: Stem and roots.

ཐྱེ་བླ།

Ngön-bhoo - **Cyananthus lobatus**
Taste: Sweet, astringent, and acrid
Potency: Cooling
Action and Use: Laxative, dries up serous fluids. Used against various serous disorders and constipation.
Parts Used: Flowers.

དངུལ་ཏིག

*Ngül-thee*k - **Parnassia cabulica**
Taste: Bitter
Potency: Cooling
Action and Use: Anti-inflammatory, cholagogue. Used against inflammation and fever from *mKhris-pa* and various contagious infections.
Parts Used: Entire plant.

ཉ་ལོ།

Nyah-loh - **Polygonum sp.**
Taste: Sour and bitter
Potency: Cooling
Action and Use: Febrifuge, anti-inflammatory, antidiarrheal. Used against inflammation of intestines, diarrhea, pain in the kidneys, hips, and lower intestines, specially after parturition.
Parts Used: Roots.

ཉེ་ཤིང་།

Nyay-shing - **Asparagus sp.**
Taste: Sweet and bitter
Potency: Neutral
Action and Use: Tonic, dries serous fluids, anti-inflammatory, heals lung disorders. Used against loss of bodily vigor, pain in hip/kidney, polyuria, skin eruptions/itching, and inflammation of the pulmonary region. It is also used against *Glo-gCong*.
Parts Used: Tuber.

པ་ཡག་རྩ་བ།

Pah-yak Tza-wah - **Lancea tibetica**
Taste: Sweet and bitter
Potency: Cooling
Action and Use: Heals pulmonary disorders, dries up blood/pus, and joins broken *Tza*. Used against pus in the lungs, cough, broken *Tza*, wounds, and coagulation of blood mixed with pus. The fruit is used for heart disorders and retention of menses while the leaves are used for healing wounds.
Parts Used: Flower, leaves, and fruit.

སྤང་རྒྱན།

P'ang-ghen - **Gentiana tubiflora**
Taste: Bitter
Potency: Cooling
Use & Action: Anti-inflammatory, febrifuge, and antitoxin. Used for redness of eyes and headache, inflammation of the throat, inflammation of the gall-bladder giving rise to yellowish skin.
Parts Used: Entire plant.

སྤང་ཙན་སྤུ་རུ།

P'ang-tsen Phoo-rhoo - **Eriophyton wallichii**
Taste: Sweet
Potency: Cooling
Action and Use: Anti-inflammatory, febrifuge, dries up pus, and joins cut nerves/vessels. Used against inflammation of the lungs, pus in lungs, inflammation of wounds/sores, and cut nerves/blood vessels.
Parts Used: Entire plant.

ར་མཉེ།

Rah-nyay - Polygonatum cirrhifolium
Taste: Sweet
Potency: Neutral
Action and Use: Tonic, promotes bodily heat, dries up serous fluids, carminative, antitussive. Used against loss of vigor, pain in the kidneys and hips, swelling and fullness in abdominal region, accumulation of fluids in bone joints, skin eruptions, cough, and *gCong* disorders.
Parts Used: Rhizome.

Rhee-sho **- Ligularia amplexicaulis**
Taste: Astringent
Potency: Cooling
Action and Use: Emetic, digestive. Used against emesis from indigestion and *Pho-rLung*.
Parts Used: Stem, leaves, and flowers.

ཪུ་རྟ།

Rhoo-thah - **Saussurea costus**
Taste: Acrid, sweet, and bitter
Potency: Neutral
Action and Use: *rLung* and blood regulator, carminative, emmenagogue, antiseptic. Used against *Khrag-rLung*, swelling and fullness of stomach, blockage and irregular menses, pulmonary disorders, difficulty in swallowing, and rotting/wasting of muscle tissues.
Parts Used: Rhizome.

གཟའ་བདུད་མགོ་དགུ   Prof. H. Harrer

Sah-dü Goh-ghoo - **Saussurea obvallata**
Taste: Bitter
Potency: Promotes heat
Action and Use: Used against *rTza* disorders, paralysis of the limbs and cerebral ischemia.
Parts Used: Entire plant.

གསེར་གྱི་མེ་ཏོག

Sayr-ghee May-thock - **Herpetospermum pendunculosum**
Taste: Bitter
Potency: Cooling
Action and Use: Febrifuge, anti-inflammatory, choleretic, cholagogue. Used against excess of *mKhris-pa*, piles, inflammation of the stomach and intestines.
Parts Used: Seeds.

སྡུད་ཤེར།

Seh-sayr - **Astragalus floridus**
Taste: Sweet
Potency: Promotes heat
Action: Tonic, diuretic, antihydrotic, emmenagogue. Used against body weakness from prolonged illness, renal inflammation from lack of exercise, lack of appetite, blood, *sKya-babs*, excessive perspiration specially when asleep, diabetes, boils/sores, diarrhea, irregular menses, and vaginal/seminal discharge.
Parts Used: Entire plant.

ཤྲི་ཁནད།

***Shree-khen-drah* - Euphorbia royleana**
Taste: Astringent and sweet
Potency: Cooling
Action and Use: Purgative, antiseptic. Used against rotting of old sores/wounds, chronic febrile, and serous disorders.
Parts Used: Latex.

ཤང་ཏྲིལ་དཀར་པོ།

Sh'ang-dreel Kar-po - **Primula involucrata**
Taste: Sweet and bitter
Potency: Cooling
Action and Use: Febrifuge, anti-inflammatory, antidysenteric. Used against contagious infections, 'hot' *rLung*, and dysentery.
Parts Used: Entire plant.

ཤང་དྲིལ་དམར་པོ།

Sh'ang-drheel Mahr-poh - **Primula macrophylla**
Taste: Bitter
Potency: Cooling
Action and Use: Anti-inflammatory, febrifuge, antidiarrheal, drains serous fluids. Used against diarrhea, inflammation of liver, gallbladder, stomach, and intestines. Especially used for children with high fever and diarrhea.
Parts Used: Entire plant.

སོ་ལོ་དམར་པོ།

***Soh-loh Mhar-poh* - Rhodiola sp.**
Taste: Sweet and slightly acrid
Potency: Cooling
Action and Use: Febrifuge, antitussive, removes odor from pulmonary tract. Used against inflammation of the lungs, common flu, bad breath, and body odor specially from the armpits.
Parts Used: Rhizome.

སུག་པ།

Sook-pah - **Silene nigrescens**
Taste: Bitter and acrid
Potency: Neutral
Action and Use: Promotes hearing, clears obstructions. Used against hard of hearing, blockage of auditory canal, and entwined intestines.
Parts Used: Flowers.

ཟུར་ལྗགས་བོང་དཀར།

Soor-lhook Bhong-kar - **Aconitum violaceum**
Taste: Bitter
Potency: Cooling
Action and Use: Antitoxin, anti-inflammatory, febrifuge. Used against poisoning from snake and scorpion bites, contagious infections, 'hot' *mKhris-pa*, and inflammation of the intestines.
Parts Used: Entire plant.

སུབ་ཀ།

Soup-kah - **Anemone rivularis**
Taste: Bitter and acrid
Potency: Promotes heat
Action and Use: Promotes bodily heat, heals 'cold' tumors,[15] antitoxin, analgesic. Used against rotting tissues, draws out serous fluids, snake poisoning, and stomachal/intestinal pain from worm infestation.
Parts Used: Seeds.

སྟག་མ།

Tagh-Mah - **Rhododendron arboreum**
Taste: Bitter
Potency: Neutral and poisonous
Action and Use: Hemostatic. Used against spreading of blood and pus in the thoracic region, specially the lungs.
Parts Used: Flowers.

ཐ་ཏྲིང་ཚང་ཤོག

***Tha-trhing Ch'ang-shock* - Pedicularis sp.**
Taste: Sweet and astringent
Potency: Promotes heat
Action and Use: Removes excess body fluids, diuretic, analgesic, and regulates respiration. Used against retention or accumulation of fluids, difficulty in micturation, breathlessness, inflammation of bone/marrow, pain from inflammation, and accumulation of serous fluids.
Parts Used: Entire plant.

སྱར་བུ།

Tahr-bhoo - **Hippophae tibetana**
Taste: Sour
Potency: Neutralizing and sharpening
Action and Use: Antitussive, expectorant, heals *sMug-po*, blood purifier. Used against disorders of the lungs, inflammation of pulmonary tract, difficulty in expelling phlegm, *sMug-po* of stomach, coagulation and clotting of blood.
Parts Used: Fruit.

ཐར་ནུ།

***Ther-nhoo* - Euphorbia cognata**
Taste: Acrid
Potency: Promotes heat and poisonous
Action and Use: Laxative, diuretic, anti-inflammatory. Used against edema, inflammatory disorders such as *Gag-Lhog*, pimples and fungal skin infections.
Parts Used: Flowers.

ཐང་ཕྲོམ་དཀར་པོ།

Th'ang-trhom Kar-poh - **Nicandra physalodes**
Taste: Acrid
Potency: Cooling and very poisonous
Action and Use: Febrifuge, anti-inflammatory, antibacterial, anthelmintic, analgesic, and increases bodily vigor. Used against contagious disorders such as *Gag-Lhog*, toothache, intestinal pain from worms, and impotence.
Parts Used: Seeds.

ཐང་ཕྲོམ་ལང་ཐང་ཙེ།

Th'ang-trhom L'ang-th'ang-tzey - **Hyoscyamus niger**
Taste: Bitter and acrid
Potency: Neutral and poisonous
Action and Use: Febrifuge, anthelmintic, dissolves tumors. Used against stomachal/intestinal pain from worm infestation, toothache, inflammation of pulmonary region, and tumors.
Parts Used: Seeds.

ཐང་ཕྲོམ་ནག་པོ།

Th'ang-trhom Nhuck-poh - **Datura stramonium**
Taste: Bitter and acrid
Potency: Cooling and very poisonous
Action and Use: Anti-inflammatory, analgesic, anthelmintic. Used against stomachal and intestinal pain from worm infestation, toothache, and fever from inflammations.
Parts Used: Seeds.

ཕུབ།

Trah-wàh - **Leontopodium sp.**
Taste: Astringent
Potency: Neutral
Action and Use: Anti-inflammatory, antitoxin, coagulant. Used primarily for moxibustion and also for contagious infections, mineral/rock poisoning, excessive loss of blood, and *rMen-bu* disorder.
Parts Used: Stem and Flowers.

སྐྱ་བཟང་།

Trah-z'ang - **Corydalis govaniana**
Taste: Sweet and bitter
Potency: Cooling
Action and Use: Antitoxin, febrifuge, anti-inflammatory, vermifuge. Used against disorders from poisoning, pain from *rLung*, swelling of the limbs, and stomachal/intestinal pain from worm infestation.
Parts Used: Entire plant.

Trhoo-nhuck - **Heracleum lallii**
Taste: Bitter and acrid
Potency: Neutral
Action and Use: Anti-inflammatory, analgesic, anthelmintic. Used against contagious disorders such as *Gag-Lhog*, swelling/pain in the joints due to '*Bam* and arthritis. Moreover, it is used against all types of pain, toothache, and inability to micturate or defecate.
Parts Used: Roots.

ཙེར་སྔོན། Prof. H. Harrer

Tsayr-ngön - **Meconopsis aculeata**
Taste: Bitter
Potency: Cooling
Action and Use: Febrifuge, analgesic, heals bone fractures. Used against inflammation from fractured or broken bones and pain in upper bodily region specially around the ribs.
Parts Used: Entire plant.

མཚེ་ལྡུམ།

Tse-Dhoom - **Ephedra gerardiana**
Taste: Bitter
Potency: Cooling
Action and Use: Stops bleeding, febrifuge, tonic. Used against loss of blood, 'hot' *mKhris-pa*, 'unripened' and 'chronic' fever.[16]
Parts Used: Stem.

ཙི་ཏྲ་ཀ།

Tzee-trah-kah - **Plumbago zeylanica**
Taste: Acrid
Potency: Promotes heat
Action and Use: Promotes heat, stomachic, digestive, vermifuge, destroys tumors, subsides piles. Used against lack of *Me-drod*, indigestion, worms and other foreign organisms in the stomach/intestines, piles, fluid accumulation, and leprosy.
Parts Used: Stem.

ཙད་གོད།

Tzeh-göh **- Pleurospermum condollei**
Taste: Bitter
Potency: Cooling
Action and Use: Antitoxin, febrifuge. Used against fever from poisoning and spreading fever.
Parts Used: Entire plant.

ཙན་དན་དཀར་པོ།

Tzen-then Kar-poh - **Santalum album**
Taste : Astringent
Potency: Cooling
Action and Use: Febrifuge, Antitussive, anti-inflammatory. Used against inflammation of lungs, heart, muscle tissues, and skin. Moreover, it is used against 'general,' 'spreading,' 'empty,' and 'turbid' fever.[17]
Parts Used: Wood and oil.

བཙོད།

***Tzöh* - Rubia cordifolia**
Taste: Bitter
Potency: Cooling
Action and Use: Febrifuge. Used against blood disorders and spreading fever of kidneys and intestines.
Parts Used: Stem.

ཚུམ་བུ།

Womb-bhoo - **Myricaria squamosa**
Taste: Astringent
Potency: Cooling
Action and Use: Localizes poison, febrifuge, ripens pimples, antitussive, dries up serous fluids. Used against inflammation from poisoning, spreading of fever from various infections, pimples that do not ripen, coughing, accumulation of serous fluids in bone joints, and meat poisoning.
Parts Used: Entire plant.

དབང་ལག

W'ang-luck **- Dactylorhiza sp.**
Taste: Sweet
Potency : Promotes heat
Action and Use: Tonic, aphrodisiac, increases semen. Used against loss of bodily vigor and sexual appetite.
Parts Used: Tuber.

གཡའ་ཀྱི་མ།

Yah-kyee-mah - **Chrysosplenium carnosum**
Taste: Bitter
Potency: Cooling
Action and Use: Anti-inflammatory, febrifuge, cholagogue. Used against *mKhris-pa*, headaches, and inflammation of the gall-bladder.
Parts Used: Entire plant.

ཡུ་མུ་མདེའུ་འབྱིན།

You-moo-dhew-chen - **Meconopsis sp.**
Taste: Bitter
Potency: Cooling
Action and Use: Opens blockages, unwinds twisted and wrapped material. Used against blockage of the menses, fetus/placenta, and removes foreign material from body such as stones, or bullets.
Parts Used: Entire plant.

ཛཱ་ཏི།

Zah-thee - **Myristica fragrans**
Taste: Acrid
Potency: Promotes heat
Action and Use: Balances *rLung*, subsides heart disorders, heatening, digestive, calmative. Used against all types of *rLung* (specially 'cold' *rLung*[18]); poor appetite, indigestion, loss of heat in the stomach and liver.
Parts Used: Fruit and oil.

ཟང་ཐིག

Z'ang-theek - **Gentian sp.**
Taste: Bitter
Potency: Cooling
Action and Use: Febrifuge, cholagogue, subsides *rLung*. Used against a combination of *rLung* and 'hot' *mKhris-pa* disorder.
Parts Used: Entire plant.

Notes

1. Since this pocketbook is primarily intended for the lay person, no specialized system of Tibetan pronunciation has been used. Thus, all Tibetan words should be pron-ounced as close as possible to the English language. However, it should be noted that all the Tibetan words such as 'P'ang,' 'Th'ang,' 'Sh'ang' and so on, should be pronounced in the same manner as hung—the past tense of hang—as in T'ang Dynasty.

2. Here, only the heating, cooling, and poisonous potencies are mainly given. On the other hand, Tibetan medicine classifies the power or potency of a drug into two main divisions. The first is the potency of the five main proto-elements of Fire, Water, Earth, Air, and Space which are directly responsible for the six tastes (sweet, sour, bitter, acrid, saline, astringent), the three post-digestive tastes (sweet, saline, bitter), the eight main drug actions (heavi-ness, greasing, cooling, blunting and their opposing actions of lightening, coarsening, promoting heat and sharpening), and the seventeen qualities of drug action such as soothing, binding, drying, moistening and so on. (See *rGyud-bZhi,* pp. 65-66).

   The second division is the actual intrinsic power of a drug which again consists of eight classifications such as aromatic power, shape of drug, auspicious power, power from prayers and so on.

3. Under no circumstances should the reader use these plants unless under the guidance of a trained physician. Tibetan medicine considers a single plant very toxic and almost all drug formulations are balanced with other plants to counteract their toxicity or to increase their potency. Thus, most Tibetan herbal drugs have a minimum of three to four plants while the more sophisticated ones have over hundred and thirty ingredients!

4. Also known as *bTsan-dug* and *Bi-sha*.

5. Ingestion of chilies, mutton, fish, chicken are examples of hot or warm diet while excessive exercise/sexual intercourse, and wearing woolen clothes in summer are examples of 'hot' lifestyle.

6. *Dha-lee* refers to the flower of *Bha-loo* (*Rhododendron anthopogon var. hypenanthum*).

7. Tibetan medicine recognizes twenty different types of tumors. However, they are generally divided into a 'hot' and 'cold' category. *mKhris-pa* and blood are the main causes of 'hot' tumors while *Bad-kan* and *rLung* are the chief causes of 'cold' tumors. For further details, see *rGyud-bZhi*, pp. 152-161.

8. Refers to a combined disorder of *Bad-kan* and *rLung*.

9. The following are some of the main *rLung* openings: posterior fontanel, two temporal regions, the palms, the soles of feet, first and sixth cervical

vertebrae, and the xiphoid process.

10. Unlike modern powerful drugs which either suppresses or scatters a fever, Tibetan medicine permits a fever to ripen and take its natural course. This method gets to the root of the disorder and invariably, a fever will be treated with the prescription of a gentle heating decoction to hasten this process of ripening.

11. *rLung* mixed with *mKhris-pa*.

12. See note 10.

13. Tibetan Buddhist medicine recognizes the following eight subtle types of fever: (1) 'unripened,' (2) 'spreading,' (3) 'empty,' (4) 'hidden,' (5) 'chronic,' (6) 'turbid,' (7) 'general,' and (8) 'borderline of mountain and plain,' i.e. a period of a disorder when there is a depletion of 'hot' blood and *mKhris-pa* and 'cold' *Bad-kan* and *rLung* are about to manifest. For further information, see *rGyud-bZhi*, pp. 177-195.

14. See note 10.

15. See note 7.

16. See note 13.

17. See note 13.

18. *rLung* mixed with *Bad-kan*.

Glossary

Bad-kan: Like *rLung* and *mKhris-pa*, this is one of the three processes of the mind/body entity. The five main types of *Bad-kan* are: (1) *rTen-byed*, (2) *Myag-byed*, (3) *Myong-byed*, (4) *Tsim-byed*, and (5) *'Byor-byed*. When they are in a state of balance and of sufficient quantity, they promote the well being of the mind/body. However, if they are in imbalance, insufficient quantity, and agitated, they give rise to 41 specific *Bad-kan* disorders. See *rGyud-bZhi*, pp. 117-128.

'Bam: A disorder where there is a proliferation of 'bad' blood which is then respectively, dispersed by *rLung*, to the muscle tissues, fats, channels, fluids, and bones. The disorder orginates from the upper region of the body and then falls to the lower limbs, particularly, the knees, and calves which become inflamed and swollen. See, Wangdu, pp. 396.

dMu-chu: A chronic fluid disorder that spreads both within and without the five vital and six vessel organs. See Wangdu, p. 436.

Gag-Lhog: A contagious disorder that affects the throat and uvula and impedes swallowing. See *rGyud-bZhi,* pp. 265-271.

gCong: A chronic metabolic disorder bringing about a

depletion of essential nutrients and loss of body weight. There are four main categories which are divided into sixteen specific types. See *rGyud-bZhi,* pp. 152-160.

***Glo-gCong*:** Another name for *gCong-chen Zad-byed* (Great Depleting *gCong*). The disorder is caused by an agitation of all body channels and openings with the result that the essential nutrients are not properly metabolized and there is the subsequent proliferation of wastes (urine, feces, and perspiration). This inevitably leads to loss of body weight, difficulty in breathing, and excessive mucal discharge from the lungs. If the disorder is not controlled in time, it can lead to various other complications. See *rGyud-bZhi,* pp. 175-177.

'Hot' *mKhris-pa*: A disorder where inflamed blood affects the gall-bladder and there is a proliferation of bile which brings about a bitter taste in mouth, excessive thirst, fever, insomnia, and yellowish feces.

***Khrag-rLung*:** A disorder where the proliferation of 'bad blood' affects *rLung* which subsequently, rises to the upper part of the body and specifically, fills up the respiratory tract and 'channels.'

***Me-bal*:** A contagious disorder that primarily affects the skin which subsequently appears to be burnt by fire. See *rGyud-bZhi,* pp. 354-355.

***mKhris-pa*:** Like *rLung,* this is one of the three main processes within the mind/body organism. The five main types of *mKhris-pa* are:(1)*'Ju-byed,* (2) *sGrub-byed,* (3) *mDangs-sGyur,* (4) *mThong-byed,* (5) *mDog-gSal*. When in a state of balance, they promote the well being of the mind/body. Otherwise, they give rise to 25 different types of *mKhris-pa* disorders. For additional information, see *rGyud-bZhi,* pp.117-125 or *Tsarong, T.J. et. al, Fundamentals of Tibetan Medicine,* pp. 46-47, Tibetan Medical Centre, Dharamsala, 1981.

Ngan-mKhrag: Because of improper metabolism within the liver, the nutritional essences (*Dangs-ma*) and the wastes (*sNyigs-ma)* are not properly separated and hence, there is a proliferation of impure blood within the blood vessels.

Pho-rLung: A *rLung* disorder that affects the stomach. Its symptoms are indigestion, eructation, emesis, stomachal rumblings, flatulence, loss of appetite, excessive thirst, and difficulty in breathing. See *rGyud-bZhi*, pp. 108.

rLung: Refers to one of the three main processes within the mind/body—the other two are *mKhris-pa* and *Badkan* —which, when in a state of balance and proper sufficiency, promotes health and well-being. If, on the other hand, there is a depletion, proliferation, or an agitation of any one of these three processes, then the mind/body entity is said to be in a state of disorder or ill-health.

All in all, there are five main types of *rLung* : (1) *Srog-'Zhin*, (2) *Gyen-rGyu,* (3) *Khyab-byed,* (4) *Me-mNyam,* and (5) *Thur-sel*. These reside at the five main energy centres of the body and when in a state of balance— both in a state of equilibrium as well as in sufficiency— are responsible for the proper mental and physiological functioning of the mind/body. If, on the other hand, this state of balance is disrupted through improper diet, lifestyle, climatic, and harmful influences, they then give rise to 20 different types of *rLung* disorders.

Incidentally, *rLung* is the most prevalent of all disorders and yet, it is ironic to note that modern medicine does not recognize this particular disorder simply because it is not reducible to matter. For additional information on *rLung*, see Yuthog Yonten Gompo et al. *rGyud-bZhi*, pp. 105-116, and T.J. Tsarong et al. *Fundamentals of Tibetan Medicine*, Dharamsala, Tibetan Medical Center, 1981, pp. 45-46.

***rLung Thur-sel*:** One of the five main types of *rLung* that is located in the perineal region. Its pathway is through the large intestine, urinary bladder, thighs, and the sexual organs. It is responsible for defecation, urination, discharging of semen/menstrual blood, and uterine contractions during parturition. For further details, see Sangye Gyatso,*Vaidurya mNgon-po*, pp. 420-451.

***rMen-bu*:** Lymphatic disorders. See *rGyud-bZhi*, pp. 356-357.

***rTza*:** Refers to all the channels and openings through which *rLung* and blood circulate. The 'white channel' (*rTza-dKar*) grows from the brain and refers to the nerves while the 'black channel' (*rTza-nag*) springs from the 'life channel' (*Srog-rTza*) i.e. the heart. (See Wangdu, *gSo-ba Rig-pa'i Tshig-mDzod gYu-thog dGongs-rGyen,* Peoples Publishing House, Beijing, 1982), pp. 463-467.

***sKya-rBab*:** This is one of the six different types of '*gCong*' disorders. According to the *rGyud bZhi,* due to improper metabolism, the nutritional essences from ingested food-stuffs and beverages, processed primarily by the liver, are not able to convert it into 'essential nutrients' (*Zung*) that sustain the mind/body. Hence, there is a proliferation of 'bad blood' and serous fluids that are dispersed by *rLung* affecting, specifically, the muscle tissues and the skin which subsequently, become swollen and whitish in colour. For additional information, see Wangdu, pp. 28-29.

sMug-po: This refers to a disorder known as 'Brown *Bad-kan.*' It is a combined disorder of *rLung, mKhris-pa, Bad-kan,* blood, and serous fluids. The 'brown' refers to a combination of colours since the natural colour of *rLung* is blue, *mKhris-pa* is yellow, *Bad-kan* is white, blood is red, and serous fluids are yellow. The original cause of this disorder is the proliferation of the Earth and Water proto-elements that subsequently, bring about lack of stomachal heat and indigestion. This, in turn, leads to improper metabolism and a proliferation of 'bad blood' and serous fluids. For additional information, see *rGyud-bZhi*, pp. 133-146.

Me-drod **(stomachal heat):** Stomachal heat is responsible for proper metabolism of ingested foodstuffs. *Bad-kan Myag-byed, mKhris-pa 'Ju-byed, rLung Me-mNyam* are primarily responsible for this function and thus, one of the main methods of Tibetan healing is to ensure the proper functioning of this heat.

sNying-rLung: It's symptoms are shivering, fullness of upper bodily region, lack of concentration, nonsensical speech, insomnia, headache, vertigo, and disorders of the skin. For more information, see *rGyud-bzhi*, p. 108.

Zung-khrag: Essential blood that is formed by the essential nutrients ingested and metabolized by the body for sustenance.

References

Deumar, T. Phuntsog, *Dud-rTzi sMan-gyi-rNam-dBye Ngo-bo Nus Ming rGyas par bShapa Shel-gong Shel-phreng,* Lhasa: Chagpori press, 18 century.

Polunin, Oleg. and Adam Stainton. *Flowers of the Himalaya.* New Delhi: Oxford University Press, 1984.

Polunin, Oleg. and Adam Stainton. *Concise Flowers of the Himalaya.* New Delhi: Oxford University Press, 1987.

Stainton, A. *Flowers of the Himalaya, A Supplement.* New Delhi: Oxford University Press, 1988.

Tsarong, T. Jigme. *Handbook of Traditional Tibetan Drugs: Their Nomenclature, Composition, Use, and Dosage.* Kalimpong: Tibetan Medical Publications: 1986.

Tsarong, T. J., et al. *Fundamentals of Tibetan Medicine.* Dharamsala: Tibetan Medical Centre, 1981.

Wangdu. *gSo-ba Rig-pa'i Tsig-mDzod gYu-thog dGongs-rGyan.* Beijing: Peoples Press, 1982.

Yuthog, Y. Gonpo., et al. *rGyud-bZhi.* Tsongon: Peoples Press, 1980.

Doland's Pocket Medical Dictionary: 22nd ed. Philadelphia: W.B. Saunders, 1977.

Health Department of Autonomous Region of Tibet. *Bod-lJong rGyun-sPyod Krung-dByi'i sMan-rigs*. Beijing: Peoples Press, 1973.

Research Department of Mentzi Khang. *gSo-Rig sNying-bsDus sKya-reng gSar-pa*. Lhasa: Peoples Press, 1978.

Index of Latin Names

Aconitum balfouri	16
Aconitum heterophyllum	15
Aconitum violaceum	84
Althea rosea	41
Amomum sabulatum	45
Androsace sarmentosa	38
Anemone rivularis	85
Arisaema sp.	29
Asparagus sp.	69
Aster diplostephioides	53
Astragalus floridus	78
Azadirachta indica	62
Carum carvi	40
Chrysosplenium carnosum	104
Clematis sp.	35
Clematis tibetana	36
Codonopsis nervosa	52
Corydalis govaniana	94
Cremanthodium sp.	60, 61
Cuscuta europaea var. indica	34
Cyananthus lobatus	66
Dactylorhiza sp.	103
Datura stramonium	92
Delphinium sp.	43
Embilica officinalis	49
Ephedra gerardiana	97
Epilobium latifolium	44
Eriophyton wallichii	72

Euphorbia cognata	89
Euphorbia royleana	79
Fragaria sp.	27
Gentian sp.	107
Gentiana sp.	23
Gentiana tubiflora	71
Geranium pratense	51
Hackelia uncinata	63
Halenia elliptica	24
Halerpestes sarmentosa	21
Hedychium spicatum	37
Heracleum lallii	95
Herpetospermum pendunculosum	77
Hippophae tibetana	88
Hyoscyamus niger	91
Inula racemosa	58
Iris Kemaonensis	31, 32
Lagotis sp.	42
Lancea tibetica	70
Leontopodium sp.	93
Ligularia amplexicaulis	74
Malva verticillata	17
Meconopsis aculeata	96
Meconopsis sp.	105
Morina longifolia	25
Myricaria squamosa	102
Myristica fragrans	106
Nicandra physalodes	90
Onosma sp.	33
Orobanche sp.	64
Parnassia cabulica	67
Pedicularis bicornuta	54
Pedicularis longiflora var. tubiformis	56
Pedicularis oliverian	55

Pedicularis pyramidata	57
Pedicularis sp.	87
Phlogacanthus pubinervius	14
Pleurospermum condollei	99
Plumbago zeylanica	98
Polygonatum cirrhifolium	73
Polygonum sp.	68
Primula involucrata	80
Primula macrophylla	81
Ranunculus acris	18
Rheum nobile	20
Rhodiola sp.	82
Rhododendron anthopogon var. hypenanthum	13, 28
Rhododendron arboreum	86
Ricinus communis	30
Rubia cordifolia	101
Rubus ellipticus	39
Rubus sp.	46
Santalum album	100
Saussurea costus	75
Saussurea graminifolia	59
Saussurea obvallata	76
Saussurea roylei	48
Saxifraga sp.	19
Silene nigrescens	83
Swertia petiolata	22
Taraxacum tibeticum	47
Thalictrum sp.	65
Thlaspi arvense	26
Tinospora cordifolia	50

Index of Tibetan Phonetics

Bah-lhoo	13
Bha-shah-kah	14
Bhong-ngah Kar-poh	15
Bhong-ngah Nhuck-poh	16
Cham-pah	17
Chay-tsah	18
Chen-zeek Tagh-nghoo	19
Choo-mah-tzee	20
Choo-rhook Bell-luck	21
Chuck-theek Kar-poh	22
Chuck-theek Nhuck-poh	23
Chuck-theek Rah-goh-mah	24
Ch'ang-tser Nhuck-poh	25
Day-gah	26
Dee-tah Sah-zheen	27
Dhah-lhee	28
Dhah-wah	29
Dhen-rock	30
Drhay-mah	31
Drhay-mare Gay-sahr	32
Drhee-mock	33
Drhül-shuck	34
E-mong Kar-poh	35
E-mong Nhuck-poh	36
Gah-kyah	37
Gah-theek	38
Gah-trah	39
Goh-nyöd	40
Hah-loh	41

Hong-lhen	42
Jah-k'ang	43
Jahr-phen Chu-tzey	44
Kah-koh-la	45
Khen-drah Kah-rhee	46
Khoor-m'ang	47
Khön-pah Ghab-keh	48
Kyou-rhoo-rah	49
Ley-trey	50
Lhee-gah-dhoor	51
Lhoo-dü Dhor-jay	52
Lhook-meek	53
Lhook-rhoo Kar-poh	54
Lhook-rhoo Mhook-poh	55
Lhook-rhoo Sayr-poh	56
L'ang-nah	57
Mah-nhoo	58
May-thock Gh'ang-lhah	59
Ming-chen Nhuck-poh	60
Ming-chen Sayr-poh	61
Neem-pah	62
Neh-mah Yhoo-loh	63
Ngoh Droh-sh'ang-tzey	64
Ngoh-trheen	65
Ngön-bhoo	66
Ngül-theek	67
Nyah-loh	68
Nyay-shing	69
Pah-yak Tza-wah	70
P'ang-ghen	71
P'ang-tsen Phoo-rhoo	72
Rah-nyay	73
Rhee-sho	74

Rhoo-tah	75
Sah-dü Goh-Ghoor	76
Sayr-ghee May-thock	77
Seh-sayr	78
Shree-khen-drah	79
Sh'ang-dreel Kar-poh	80
Sh'ang-dreel Mahr-poh	81
Soh-loh Mahr-poh	82
Sook-pah	83
Soor-lhook Bhong-kar	84
Soup-kah	85
Tagh-mah	86
Tha-trhing Ch'ang-shock	87
Tahr-bhoo	88
Ther-nhoo	89
Th'ang-trhom Kar-poh	90
Th'ang-trhom L'ang-th'ang-tzey	91
Th'ang-trhom Nhuck-poh	92
Trah-wah	93
Trah-z'ang	94
Trhoo-nhuck	95
Tsayr-ngön	96
Tse-dhoom	97
Tzee-trah-kah	98
Tzeh-göh	99
Tzen-dhen Kar-poh	100
Tzöh	101
Womb-bhoo	102
W'ang-luck	103
Yah-kyee-mah	104
You-moo Dhew-chen	105
Zah-thee	106
Z'ang-theek	107

Index of Tibetan Names

ཀ་གོ་ལ།	45
གནད་ཀ་རི།	46
གོན་པ་སྐྱབ་སྐྱེས།	48
སྒྲུ་རུ་ར།	49
སྒྱུ་བདུད་རྡོ་རྗེ།	52
སྒྲུ་བཟང་།	94
ཁུར་མང་།	47
ག་བྲ།	39
གོ་སྙོད།	40
ཤྱང་སྨུག	57
སྨ་དིག	38
སྨ་སྒྲུ།	37
དངུལ་དིག	67
རྡོ་སྙིན།	65
རྡོ་བྲ།	66
རྡོ་སྦོ་ནག་ཙེ།	64
བཙན་ཞིག་ཧག་ཏུ།	19

ལུགས་ཉིག་ར་མགོ་མ།	24
ལུམ་པ།	17
ལྱེ་ཚ།	18
ལུགས་ཉིགས་དཀར་པོ།	22
ལུགས་ཉིགས་ནག་པོ།	23
ཆུ་མ་རྩི།	20
ཆུ་རུག་སྒལ་ལག	21
ཉེ་ཤིང་།	69
སྐྱུ་ལོ།	68
སྦག་མ།	86
སྦར་བྲུ།	88
ཐང་ཕྲོམ་དཀར་པོ།	90
ཐང་ཕྲོམ་ནག་པོ།	92
ཐང་ཕྲོམ་ལང་ཐང་རྩེ།	91
ཐ་རྡིང་ཅང་ཤོག	87
ཐར་བུ།	89
དུ་ཡིས།	28

དུབ།	29
དན་རོག	30
དྲེས་མ།	31
དྲེས་མའི་གི་སར།	32
ནད་མ་གཡུ་ལོ།	63
ནིམ་པ།	62
པ་ཡག་རྩ་བ།	70
སྤང་རྒྱན།	71
སྤང་ཚན་སྔུ་ཏི།	72
སྤང་ཙེར་ནག་པོ།	25
སྤུ་བ།	93
སྤུ་ནག	95
བ་ལུ།	13
བ་ཤ་ཀ།	14
བོང་ང་དཀར་པོ།	15
བོང་ང་ནག་པོ།	16
བུ་ཤུད།	43

བྲི་ག	26
དབང་ལག	103
དབྱི་མོང་དཀར་པོ།	35
དབྱི་མོང་ནག་པོ།	36
བུར་པན་ཆུ་རྩི།	44
སྒུལ་ཞགས།	34
འབྲི་ཏ་ས་འཛིན།	27
འབྲི་མོག	33
མ་ནུ།	58
མེ་ཏོག་གངས་ལྷ།	59
མིང་ཅན་ནག་པོ།	60
མིང་ཅན་སེར་པོ།	61
ཙན་དན་དཀར་པོ།	100
ཙོ་ཏྲ་ཀ།	98
བཙོད།	101
ཚད་ནོད།	99
ཚེར་སྔོན།	96

མཚོ་ལྷམ།	97
ཇ་དི།	106
ཟང་ཏིག	107
ཟུར་ལུགས་བོང་དཀར།	84
གཟའ་བདུད་མགོ་དགུ	76
འུམ་བུ།	102
ཡུ་མུ་མདེའུ་འབྲིན།	105
གཡའ་ཀྱི་མ།	104
ར་མཉེ།	73
རི་ཤོ།	74
རུ་རྟ།	75
ལི་ག་དུར།	51
ལུག་མིག	53
ལུག་རུ་སེར་པོ།	56
ལུག་རུ་དཀར་པོ།	54
ལུག་རུ་སྨུག་པོ།	55
ཤང་དྲིལ་དཀར་པོ།	80

ཤང་རྒྱལ་དམར་པོ།	81
ཤྲི་ཁནད།	79
ཤུག་པ།	83
ཤུབ་ཀ།	85
ཤུད་ཤིར།	78
ཤོ་ལོ་དམར་པོ།	82
སྐྱི་ཏྲིས།	50
གཤེར་གྱི་མེ་ཏོག	77
ད་ལོ།	41
ཏོང་ཞིན།	42